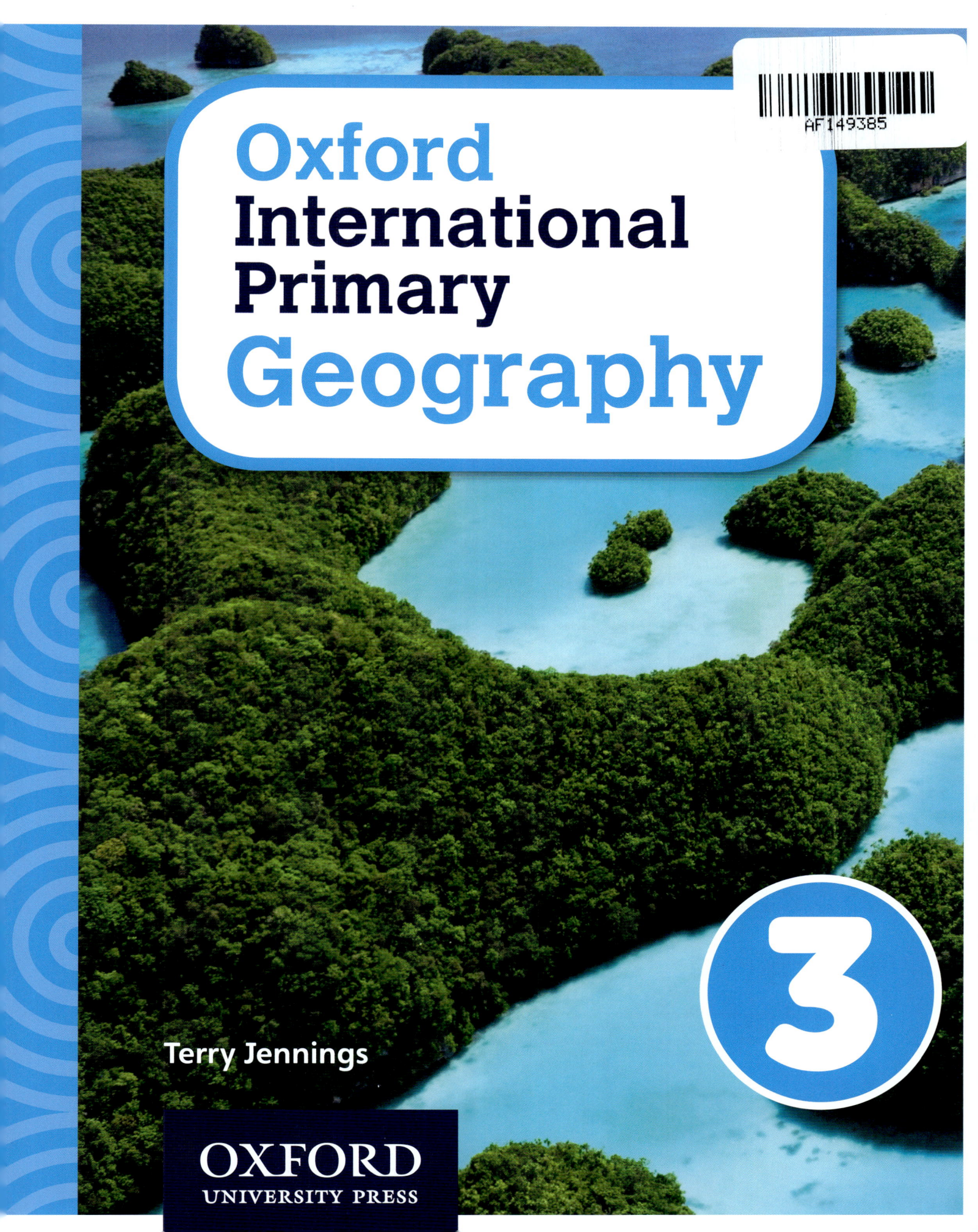

Oxford International Primary Geography

3

Terry Jennings

OXFORD
UNIVERSITY PRESS

Great Clarendon Street, Oxford, OX2 6DP, United Kingdom

Oxford University Press is a department of the University of Oxford. It furthers the University's objective of excellence in research, scholarship, and education by publishing worldwide. Oxford is a registered trade mark of Oxford University Press in the UK and in certain other countries

British Library Cataloguing in Publication Data
Data available

978-0-19-831005-1

20 19

The manufacturing process conforms to the environmental regulations of the country of origin.

Printed in Great Britain by Bell and Bain Ltd, Glasgow

Acknowledgements

The publishers would like to thank the following for permissions to use their photographs:

Cover photo: Getty Images/Reinhard Dirscherl, P6a: Dream Perfection/Shutterstock, P6b: © Phil Degginger / Alamy, P8a: © Chris Hellier/Corbis, P8b: © Paul Panayiotou/Paul Panayiotou/Corbis, P8c: © LOOK Die Bildagentur der Fotografen GmbH / Alamy, P9: © Yadid Levy/Robert Harding World Imagery/Corbis, P10: © Yadid Levy / Alamy, P11: © Amar Grover/JAI/Corbis, P12a: © Sasha Woolley/Corbis, P12b: Boris Stroujko/Shutterstock, P12c: © David Bathgate/Corbis, P13a: © Robert Francis/Robert Harding World Imagery/Corbis, P13b: Gerard SIOEN / Contributor/ Getty Images, P14a: Frolova_Elena/Shutterstock, P14b: Pawel Kazmierczak/Shutterstock, P15a: iStock, P15b: iStock, P16: Volodymyr Goinyk/Shutterstock, P17: © Charles Bowman/Robert Harding World Imagery/Corbis, P18: Boris Stroujko/Shutterstock, P19a: © Bildarchiv Monheim GmbH / Alamy, P19b: Günter Gräfenhain/4Corners, P20: REUTERS/Amr Abdallah Dalsh, P21a: © Jim Holmes/Design Pics/Corbis, P21b: © Bruno Morandi/Corbis, P22a: ISRAEL-PALESTINIANS/BUILDING REUTERS/Mohamad Torokman, P22b: Simon Rawles/ Getty Images, P24: © Svenja-Foto/Corbis, P25: Richard Packwood/Getty Images, P26a: Rawpixel/Shutterstock, P26b: bondgrunge/Shutterstock, P27: Steve Wilson/Shutterstock, P29: © Pacific Press/Corbis, P32a: Paul Michael Hughes/Shutterstock, P32b: iStock, P33: Denys Prykhodov/Shutterstock, P34: © Atlantide Phototravel/Corbis, P35a: © Uden Graham/Redlink/Corbis, P35b: Greg Elms/Getty Images, P36a: © mohammad asad/Demotix/Corbis, P36b: AP Photo/Bullit Marquez, P38a: © Phil Wills / Alamy, P38b: SSPL/Getty Images, P38c: Paul Nicklen/National Geographic/Getty Images, P39: © RG Images / STOCK4B/Stock4B/Corbis, P40a: © Michael Freeman/Corbis, P40b: © Ludovic Maisant/Hemis/Corbis, P41: © Zhang Chunlei/Xinhua Press/Corbis, P42a: © Keith Dannemiller/ Corbis, P42b: © Marcel Malherbe/Arcaid/Corbis, P43: © Eric Robert/Sygma/Corbis

Although we have made every effort to trace and contact all copyright holders before publication this has not been possible in all cases. If notified, the publisher will rectify any errors or omissions at the earliest opportunity.

Links to third party websites are provided by Oxford in good faith and for information only. Oxford disclaims any responsibility for the materials contained in any third party website referenced in this work.

The manufacturer's authorised representative in the EU for product safety is Oxford University Press España S.A. of El Parque Empresarial San Fernando de Henares, Avenida de Castilla, 2 – 28830 Madrid (www.oup.es/en or product.safety@oup.com). OUP España S.A. also acts as importer into Spain of products made by the manufacturer.

Contents

Maps

Imagine you are in an airplane, looking down at your school. What would it look like? A map shows the view of an area from above. We use maps to find out where places are, or how far it is between places.

Map scales

When people draw maps, they draw them much smaller than the area of land or sea the maps show. Maps are nearly always drawn to **scale**. Different scales are used to draw different types of map.

Map	Distance on paper	Actual distance on the ground
School	1 cm	10 m
Town or city	1 cm	250 m
Small country	1 cm	50 km

A large-scale map shows a small area in a lot of detail. It is the kind of map you might use if you are out walking. A small-scale map shows a large area in less detail. Who would use this type of map?

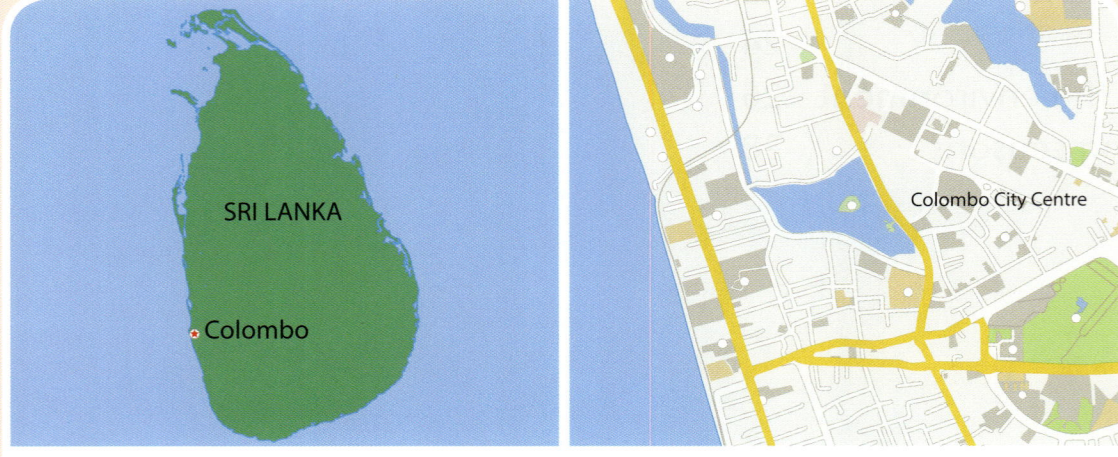

SRI LANKA

Colombo

Colombo City Centre

Which of these maps is a small-scale map? How can you tell? What things does the large-scale map show?

Globes and maps

A **globe** or the maps in an **atlas** are small-scale maps. They show huge areas, but with very little detail. Is the map on pages 44 and 45 of this book a large-scale map or a small-scale map?

Map symbols

The things that maps show are marked by **symbols**. Some symbols are tiny pictures. Others are initial letters, such as 'H' for heliport or 'P' for car park. Most maps have a **key** to tell you what these different symbols mean. There are no rules about what colours are used on maps, but map-makers use the same colours for the same things. Water is usually blue, for example, and forests are green.

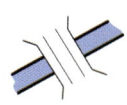     

| Bridge | Airport | Castle | Heliport | Lighthouse | Cave |

Activities

1. As a class, collect different-scaled maps of your local area. Compare them.

   a Which map shows the most detail?

   b Which map shows the largest area?

2. This is part of a bird's-eye view of a village. Draw a map of it. Will you include the car, truck and camel in your map? Say why.

3. Make a class display of the symbols used on various maps. Underneath each symbol write what it means.

Directions

Maps do more than help us to find out where places are and the distances between them. Maps help us to find the way. This is because they tell us which direction to take to reach a place.

North

Most maps have an arrow on them showing which way is north. This arrow always points towards the top of the map. To find out which direction is north from where you are, you need a **compass**.

Compasses

A compass has a needle, which is a thin magnet. The end of the needle always points to the **North Pole**. If you know which way north is then you can work out where the other directions are.

The four main points on a compass are north (N), south (S), east (E) and west (W).

With a compass and a map, you can find your way in fog or even in a sandstorm when it is hard to see where you are going.

Finding the way

To work out which direction to take, you need to use your map and your compass. Place the compass on the map and turn the map around carefully until the north arrow on the map points in the same direction as the needle on the compass.

Finding the way with a map and compass.

Did you know?

The first person to use a compass to find the way was Zheng He, a sailor from China. He made seven ocean voyages between 1405 and 1433.

Other compass points

There are four more compass points it is useful to know:

- North-east (NE) is halfway between north and east.
- South-east (SE) is halfway between south and east.
- South-west (SW) is halfway between south and west.
- North-west (NW) is halfway between north and west.

Activities

1   There are nine letters in this rectangle. Use the words north, south, east and west to say how you would travel from one letter to another. Some of the questions have more than one answer. Use the word 'then' to connect two compass points. The first question has been answered for you.

A	B	C
D	E	F
G	H	I

Which direction do you travel going from A to G?

(Answer: south to D, then south to G)

Which direction do you travel going from C to G?

Which direction do you travel going from B to I?

Which direction do you travel going from A to H?

Which direction do you travel going from E to A?

2   Copy and complete these sentences, using the four main compass directions. Use an **atlas** to help you.

Sudan is _____ of Egypt.

Bangladesh is _____ of India.

Canada is _____ of the United States of America.

Bulgaria is _____ of Romania.

Portugal is _____ of Spain

Chile is _____ of Argentina.

Villages

The groups of buildings where people live are called **settlements**. Villages, towns and cities are all kinds of settlements. What kind of settlement do you live in?

Village settlements

A village is a small settlement in the countryside. Fewer people live in a village than in a town or city and there are fewer roads and buildings. In most villages there is a place of worship. There may be a few shops. Larger villages may have a school, a garage, and a coffee shop.

Roquebrunne, in southern France was built on a hill surrounding the castle, so villagers were protected against invaders.

Many seaside villages were built on a hill or cliff near the sea. Other villages developed where there were sheltered harbours or where there was good fishing.

The tiny village of Mousehole in the UK has a sheltered harbour.

On the flatter areas of land away from the coast, or along river **valleys**, the villages are usually small and the houses are made of materials found nearby.

In **mountain** areas the houses are made of stone.

Karmi village in northern Cyprus has freshwater **springs** from the limestone rocks of the surrounding mountains.

Village squares

In many larger villages, the houses, shops and other buildings are situated around a central square, where the weekly market is held. This is often used as a place for children to play and for adults to meet and talk.

The village of Ingiriya, and its small market, grew where two roads meet in Sri Lanka.

Village roads

All villages have roads passing through them. Originally farmers needed roads to get to and from their fields. The roads were also needed to move goods and people in and out of the village. Villages often grew where two or more roads met. Nowadays, more and more villages are connected by good roads. This means that people can travel greater distances. They can quickly and easily get to work or take their produce to market in the towns and cities.

Did you know?

In South Africa the villages are called kraals. The people build a fence around each kraal to keep their animals in and to keep wild animals out.

Activities

I Work in groups and use an **atlas** or maps to find the names of ten villages in your area, or from any area that you know.

  a How is each village set out? Is it a single **street** or grouped around a square?

  b What features does each village have? Write a list for each village.

2 Do you know any villages in other countries? Use reference books or the Internet to find out what some of them are like.

The village of Dana

Dana is a small village near the city of At-Tafilah in Jordan. It is a cluster of flat-roofed houses made of stone and wood. They are plastered with mud. Dana is built on the edge of a large **gorge**, called Wadi Dana. Many people believe that it is one of Jordan's loveliest villages.

Traditional flat-roofed houses in Dana.

Ancient Dana

There have been houses on the site of Dana since ancient times. People first moved to Dana because of its three **springs**. These provided fresh water and good grazing for animals. The villagers were able to grow all kinds of fruits and vegetables.

An empty village

For most of the 20th century Dana was a farming village. As Jordan became richer, the people of Dana began to move away to find a better life in the towns and cities. In the early 1980s a large cement factory was built a short distance away. Even more people moved away from Dana to a new village near the cement factory.

Bringing life back to the village

Dana was almost empty of people for about ten years. In the early 1990s a group of people decided to try to bring life back to Dana. Electricity, water, a sewage system and telephone lines were put in. About 65 of the old houses were rebuilt, some for use by **tourists**.

People began to move back to Dana. The farmers still grew the same **crops**, but they made jams and soaps that tourists would buy. Some local people also designed and made silver jewellery. When the Dana Nature Reserve was created in the wild **mountains** on the other side of Wadi Dana there were even more jobs for local people.

Today Dana is a thriving village. The streets have been rebuilt. There are hotels and campsites for tourists, and new parks and play areas for everyone to enjoy.

The Wadi Dana Gorge attracts many tourists to the area.

Did you know?

The world's oldest city that people still live in is Arbil in the South Kurdistan region of Iraq. It is more than 8000 years old and people are still living within its walls.

Activities

1 Write two lists to compare where you live with Dana: one list of the things that are the same and the other of the things that are different.

2 Would you like to live in Dana? Give reasons for your answer.

3 In many parts of the world, villages are changing. Some villages are growing. Others are getting smaller. Write a list of all the reasons you can think of why a village might change. Think about the people, jobs, shops, roads and transport, communications, **farm** machinery and money.

Farms and food

Some of the food we eat comes from **farms** in our country. The rest is brought to us from other countries.

Some farmers produce only enough to feed their families and have little or nothing to sell. In other places huge farms sell their produce to people in other countries.

A cattle ranch in Australia. Farms like this rear thousands of cattle to sell to other countries for meat.

What crops need

Crops need water, sunshine and warmth if they are to grow well.

In most of the Middle East, there is plenty of warmth and sunshine but not enough rain for most crops. As a result, crops can be grown only where there is water. These places are called **oases**. Some oases are small villages. Other oases are towns and have important markets.

Crops also need **fertile** soils if they are to grow well. In much of India there is warmth, sunshine and rain, but the soil is not very fertile. Expensive chemical **fertilisers** are needed to make crops grow.

These date palms and other crops can grow in this oasis because there is enough water for them.

Planting rice in Bangladesh. Young rice plants need to stand in water while they are growing.

Irrigation

Another way in which crops can be grown in very dry soils is with the help of **irrigation**. Irrigation means putting water onto the land. Farmers take water from rivers, lakes or wells and let it flow through ditches or pipes onto their land.

These wheat seedlings in Vietnam are being irrigated.

Egypt is almost all desert. Most Egyptian people live near the River Nile. The country's food comes mainly from near the Nile, where fertile soils and water for irrigation make farming possible.

Most of Egypt's food crops are grown near the River Nile.

Farm chemicals and organic farms

Many farmers put fertilisers on their land and spray **pesticides** on their crops. By using chemicals farmers can produce more food, but they can damage the soil, **pollute** rivers and streams and kill wildlife.

Some farmers grow organic **crops**. They do not use any chemicals. Many people believe that the produce tastes better although it does not grow as well.

Organic farming is good for wildlife. Most supermarkets now sell organic food, so there are more organic farms.

Activities

1 Work with a friend. Talk about what you would like and dislike about working on a farm.

2 Write a list of the foods you eat for breakfast or dinner. Draw pictures of the items and then find out where the main ingredients are grown.

Choosing a holiday

What things help you to decide where and when to go on holiday? One thing to check is the **weather**.

In some places, the weather changes little all year round.

However, some parts of the world have hot and cold **seasons**, or wet and dry seasons. The weather affects what we wear, eat, drink and do.

Leisure activities

The weather affects our leisure activities. If you wanted to fly a kite or sail a boat you would choose a dry, windy day. If you wanted to have a picnic or go to the seaside, you would choose a warm, sunny day.

Weather and holidays

Many people now go to another country for their holiday. They choose a country that has the kind of weather they want for their type of holiday.

Where in the world?

What is the best weather for sitting on a beach or swimming in the sea? The countries around the Mediterranean Sea, the islands in the West Indies, the beaches of eastern Australia and the countries of the Middle East mainly have hot, sunny weather.

Kitesurfing in Hawaii.

Swimming in the Mediterranean Sea.

What is the best weather for skiing, snowboarding, or other winter sports? The mountainous parts of Europe, North America and New Zealand have cold winters with plenty of snow.

If you want to go on holiday to a big city or to look at museums and old buildings, it might not matter so much what the weather is like.

Skiing in Italy.

The Louvre Museum in Paris, France.

Activities

Help the family below choose where to go on holiday.

The Jensen family want a winter holiday in December. Mr and Mrs Jensen are keen to try out their new skis. Their 12-year-old son, Johann, wants to ski too, but he also hopes to be able to go bird-watching in forests and by lakes. Johann's 9-year-old sister, Alexa, wants to take lots of pictures with her new digital camera. They all want to try the food in the local coffee shops and restaurants.

Work in a group of three. Collect some holiday brochures or use a travel agent's website and choose three places where the family could go to on holiday. Use guidebooks or the Internet to find out more about these places, including what the weather is like.

a Record what you have found out about each place.

b How much would two weeks' holiday cost for the family at each of the three places?

c Which place do you think the family will choose? Why?

Hot and cold places

Why do different parts of the world have different kinds of **weather**? We call the usual weather of a place its **climate**.

In New Zealand and Northern Europe, the weather changes from day to day. In these places it is usually much colder in winter than in summer and it rains quite a lot all the year round. Other places have a climate that is much less changeable. Some places are very hot all the year round. Other places are very cold. In some places it rains a lot. Other places are dry. What is the climate like where you live?

The Equator

The climate of a place depends mainly on how close it is to the **Equator**. This is an imaginary line around the middle of the Earth. Near the Equator the Sun's rays shine straight down from high in the sky. The Sun warms the surface of the Earth most strongly at or near the Equator. That is where the hottest places are.

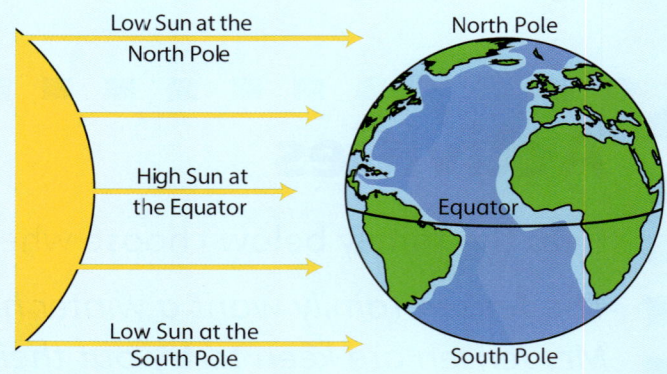

The Sun's rays are strongest over the Equator.

The polar regions

The surface of the Earth is curved. Places further away from the Equator are warmed less strongly by the Sun and so they are colder. The coldest places of all are near the top and bottom of the Earth: the **North Pole** and the **South Pole**. It is so cold that there are large areas of snow and ice.

The land around the South Pole is very cold all the year round. Which animals live around the South Pole?

Mountains and the sea

Other things also affect the climate. As you go higher up a **mountain**, for example, it gets colder. The **temperature** falls about 2°C for each 300 metres you climb. There is snow all the year round on the tops of some high mountains near the Equator.

Mount Kilimanjaro in Tanzania is Africa's highest mountain. Although it is near the Equator, its top, or summit, is always covered in snow.

The sea also affects the climate. Water holds heat better than the land does. When the weather is hot, places near to the sea are cooler than places inland, but when the weather is cold, places nearer to the sea are warmer than places inland.

Did you know?

Antarctica is the coldest place on Earth. On 21 July 1983, the temperature at the Vostok Research Station plunged to −89.2°C.

Activities

1 Look at a **globe** or a map of the world. Run your finger along the Equator.

   a Write a list of the countries the Equator passes through.

   b What large cities does the Equator pass very near to or through?

   c What do you think the climate of these places will be like? Use reference books or the Internet to find out if you were right.

2 Use an **atlas** to find which continents:

   a have large areas of very cold lands

   b have large areas of desert

   c have large areas of tropical rainforest (these are near the Equator).

Looking at Switzerland

Switzerland is a small country in Europe. About eight million people live there. Switzerland is completely surrounded by other countries. We call a country that has other countries all around it 'landlocked'.

Map of Switzerland

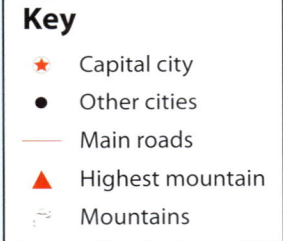

Key

★ Capital city

● Other cities

— Main roads

▲ Highest mountain

≋ Mountains

30 km

20 mi

Sisikon village

Rachel lives in Switzerland in a small village called Sisikon. About 400 people live in Sisikon. They speak German, like most Swiss people, although a few also speak Italian or French. The average maximum **temperature** in Sisikon in January is 2°C, rising to 25°C in July. Although summers in Sisikon are warm, they are fairly wet. In winter the snow lies deep almost down to the edge of the lake.

Sisikon is in a small valley on the shores of Lake Lucerne.

The Alps

The snow-capped **mountains** of Switzerland, called the Alps, make the **climate** very changeable. The Alps attract many holidaymakers in winter and summer. Long tunnels and bridges take the roads and railways through the mountains and across **valleys**.

Employment and land use

Sisikon has two shops. It also has several hotels and guesthouses, and a campsite where holidaymakers can stay. There is a main road and a railway, which carry people and goods between Italy and Switzerland and other parts of Europe.

Leisure boats regularly call in at Sisikon and take passengers to other towns and villages on Lake Lucerne.

Different types of leisure boats on Lake Lucerne, moored in Lucerne.

There are several small **farms** around Sisikon. The farmers graze their cattle on the high mountain **meadows** in summer. The farmers keep the cattle for meat and milk.

Some of the milk is made into cheese. In the winter the cattle are kept indoors and fed on **hay** that was cut from the meadows during the summer. The soil is too thin and the slopes are too steep for **crops** to be grown around Sisikon.

The ground around Sisikon is so steep that farmers use it for grazing animals rather than growing crops.

Activities

1 Use an **atlas** and find Switzerland and Lake Lucerne.

   a What are the names of the countries that surround Switzerland?

   b Which of these countries are also landlocked?

   c Use reference books or the Internet to find out what these countries are like. Write a sentence or two about each one.

   d Which large river has its **source** in Switzerland and flows north through France, Germany and the Netherlands?

2 Which of these countries have a coastline and which are landlocked?

   Australia   Austria   Bhutan   Bolivia   Botswana   Chad
   China   Italy   Japan   Mali   Paraguay   Zambia

Life in Cairo

Cairo is the **capital** city of Egypt and the largest city in Africa. Over nine million people live in Cairo. It lies on the banks of the River Nile, the longest river in the world.

Cairo is a noisy, busy city, crowded with people and traffic.

Cairo is a noisy, crowded city in which the streets are jammed with **traffic** all day long. To one side of Cairo are the narrow alleys and bazaars (markets) of the old city. On the other side are the modern office blocks, banks, hotels and wide streets of the new part of Cairo.

Cairo's climate

The **climate** of Cairo is sunny, hot and dry, although a little rain may fall in the winter. The average maximum **temperature** in Cairo in January is 19°C, rising to 35°C in July, although it is very cold at night all year round. How does this compare with Sisikon (see page 18)?

Home, work and travel

Mahmood Abdallah is almost 8 years old. He lives in a small apartment in the new part of Cairo with his parents and his 3-year-old sister Yasmine. Mahmood's father is a journalist, so he has to travel a lot. Mahmood's mother looks after the apartment and her family.

Schooling

Like most people in Egypt, Mahmood speaks Arabic. He has been going to school since he was 6 years old. Already Mahmood can read and write.

A classroom in Egypt.

Tourism

Thousands of **tourists** visit Cairo every year. Cairo has many very old mosques, museums and bazaars.

In the bazaars, people make and sell shoes, pots and pans, jewellery, gold ornaments and souvenirs.

Most tourists also visit the famous pyramids and Sphinx in Giza, about 10 kilometres south of Cairo. Many of the tourists also travel on cruise ships on the Nile.

A cruise ship on the Nile.

Did you know?

Cairo has the only metro or subway system in Africa. It is one of the world's busiest.

Activities

Look at an **atlas** and answer these questions:

a What is the name of the sea that separates Egypt from Europe?

b Roughly how far from the sea is Cairo?

c What is the name of the sea that separates Egypt from Saudi Arabia?

d What countries lie to the west and south of Egypt?

e What is the nearest country across the sea north of Egypt?

My environment

What do we mean by our **environment**?

Our environment is our surroundings. It is everything around us including the land and water, the air we breathe and the plants and animals (including people). **Mountains**, hills, **valleys** and coasts are part of our environment, as are the villages, towns and cities where we live. Some of our environment is natural, but much of it has been made or changed by people.

Changing environments

Every day we live in several environments. Your home and your classroom are two different environments. In which other environments do you live in a day?

These pictures show the view from the windows of two schools. What can you see in each view? What do you think the environment is like around each school?

The view from the window of a village school.

The view from the window of a city school.

Plants and us

Wherever we live, in a city, a town or the countryside, we depend upon plants and other living things, and they depend upon us. Green plants are the only living things on Earth that can make their own food. To do this they use the energy of sunlight, water from the soil and a gas called carbon dioxide from the air. Carbon dioxide is in the air that living things breathe out. It is also made when we burn coal, oil, gas, wood and other materials.

While they are making their food, green plants also make the gas oxygen, which all living things need to breathe. Animals, and people, cannot make their own food. They have to eat either plants, or other animals that eat plants, in order to stay alive.

Did you know?

The Earth formed about 4570 million years ago. The first animals with shells and bones appeared less than 600 million years ago.

Activities

1 Look out of your classroom window. Think about what you can see, hear and smell.

   a Write a short description of the view.

   b How might the view be different at night? Describe it.

2 What plants and animals can you see from your classroom window? What could you do to encourage more plants and animals to live there?

3 One of the things that spoils the appearance of buildings and streets is litter. Why do you think people leave litter? Draw or write some ideas about how you could improve the appearance of an area that has a lot of litter.

Using land

How is the land used around your school?

We use land in all kinds of ways. We use it to grow our food and the trees from which we get timber and paper. We also use land for buildings and for roads, railways and airports. Parks, gardens and playing fields all use land. Almost all land is used for something.

We can see how land is used if we look at an **aerial photograph** or a map.

This is part of Cyprus. What types of land use can you see?

Land use in towns and cities

A town or city covers a much larger area of land than a village. Near the centre of town or city there are shops. There may also be restaurants, hotels, theatres, cinemas and offices. As well as roads there may be a bus station or a railway station. Many towns and cities have public gardens or a park.

Further out from the town or city centre there are likely to be many houses and apartments, as well as schools, garages, factories and smaller shops. In all parts of the town or city there may be places of worship, cemeteries and car parks.

In recent years, some large supermarkets and modern factories have been built on the outskirts of towns and cities. At the very edge of a large city there may be an airport.

Did you know?

Since prehistoric times, people have been turning forests, wild grasslands, marshes and even some deserts into farmland. Now more and more farmland is being used to build houses, shops, offices, factories and roads.

Village land use

In and around a village, a great deal of land may be used to grow **crops** or to feed **farm** animals. There may be areas of trees. Land will also be used for roads, places of worship, houses, gardens, shops and coffee shops, and possibly the village school with its playground or playing fields.

This village is in Oman. What types of land use can you see?

Activities

1 Work in a group.

   a Talk about how your local area has changed in the past. Look at old maps and photographs. You could interview some people who have lived in the area for a long time.

   b How is the area changing now?
   Write and draw about what you have learned.

2 Find a large-**scale** map of your street or the area around your school. Carefully colour in the different buildings according to what they are used for. Use one colour for each type of building you choose. These could include houses and apartments, schools, shops, factories, places of worship, places of leisure, and coffee shops and restaurants. Write a **key** to your map.

3 Look around your local area. Find a piece of land where the use has changed recently. Write down the old and new uses of the land. Is the new use better or worse than the old one? Say why.

People and the environment

Almost everywhere you look there are people. People travel all over the world. These days it is difficult to go anywhere that people have not already visited.

Early people

Nobody knows for sure when the first people appeared on the Earth. They probably first appeared in Africa. When humans spread out from Africa there were between 2000 and 5000 people in the whole world. Today there are more than seven billion (seven thousand million) people and the number is still growing.

Changing environments

In the early days there were so few people and so much space for them that they made little difference to their **environment**. Nowadays, there is hardly anywhere on Earth that people have not affected.

Today, huge areas of land are cleared to grow food and to build new homes, shops, offices, factories and roads. People dig for building stone, **minerals** and some **fuels**.

This quarry is in Russia. Valuable soil has been destroyed and the appearance of the countryside has been spoiled.

We harm our environment with noise and litter. We **pollute** the air, water and soil.

Endangered species

Although there are many kinds of animals and plants in the world today, some have died out. We say they have become **extinct**. Many other kinds of plants and animals are rare, which means that only a few are left. We say they are **endangered**. These plants and animals have become endangered because:

- they lost their homes when people cleared land for farming or for building new roads, houses, shops, offices and factories

- they have been poisoned by waste chemicals from **farms** and factories

- they have been hunted for their meat, fur, skins or horns.

The tiger is endangered because of hunting and because people are destroying the forests where tigers live.

Activities

1 Work with a friend. Write a plan to improve your school grounds for wildlife. Use your own ideas and look in reference books too.

   a What special equipment would you need?

   b Who would look after the plants and animals during the school holidays?

2 Create a spreadsheet of endangered wildlife. Use the following headings: Animal or plant, Habitat or home, Numbers (if known), Threat. Use reference books and the Internet to help you to complete the spreadsheet.

The changing seasons

Our **environment** changes slowly and steadily because of the things we do. Regular changes to the environment take place every year because of the changing **seasons**.

What are the seasons?

A season is a time of the year when you can expect a special type of **weather**. In Northern Europe and New Zealand, for example, it is hotter during the summer and colder during the winter. It can rain at almost any time. These places, which are a long way north or south of the **Equator**, have the four different seasons we call spring, summer, autumn and winter.

The changing seasons affect the way people live and the **crops** they can grow. The seasons also affect the clothes we wear and the places we visit.

Different seasons

Near the Equator it is nearly always hot, although there are dry seasons and rainy seasons in some areas. Around the Mediterranean Sea it is hot and dry in summer from May to August. There is a short autumn season during September and October when the weather is changeable. This is followed by a cool and wet winter that usually lasts from November to February. There is a short spring with changeable weather during March and April.

The countries of South-East Asia, such as India, Bangladesh, Burma, Thailand and Vietnam, have what is called a **monsoon climate**. From March to June the weather is dry and hot. This is followed by heavy monsoon rains between July and October, and then dry and warm weather from November to February.

Heavy monsoon rains often bring flooding to places in South-East Asia.

The tilted Earth

One reason why we have seasons is because the Earth moves around the Sun. It makes one complete journey around the Sun every year. The Earth is slightly tilted and so the weather is different at different times of the year. The diagram shows what happens.

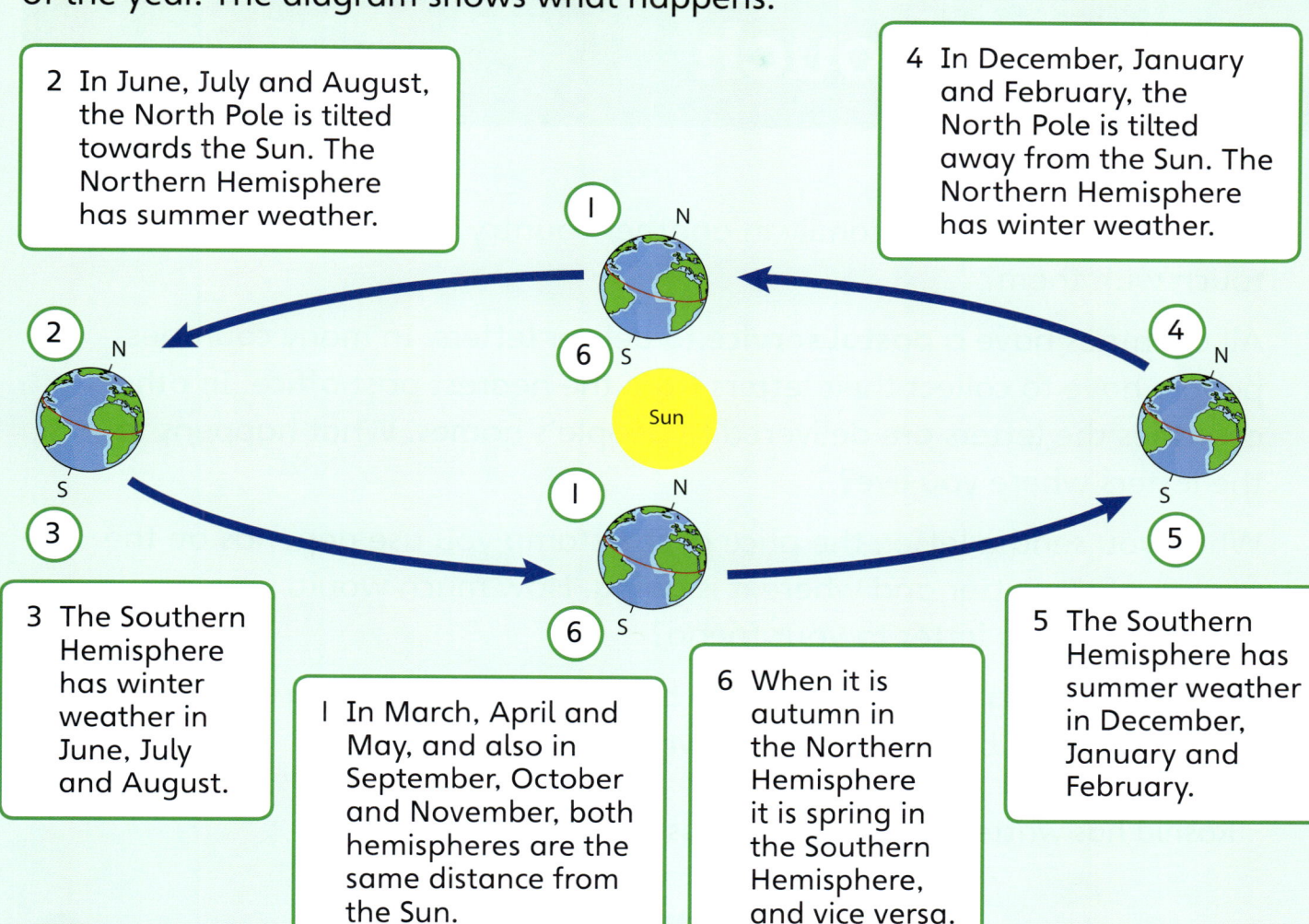

2 In June, July and August, the North Pole is tilted towards the Sun. The Northern Hemisphere has summer weather.

4 In December, January and February, the North Pole is tilted away from the Sun. The Northern Hemisphere has winter weather.

3 The Southern Hemisphere has winter weather in June, July and August.

1 In March, April and May, and also in September, October and November, both hemispheres are the same distance from the Sun.

6 When it is autumn in the Northern Hemisphere it is spring in the Southern Hemisphere, and vice versa.

5 The Southern Hemisphere has summer weather in December, January and February.

Activities

1 Imagine you live in Northern Europe or New Zealand. Write down four activities you might do in:

   a summer   **b** winter.

2 Use reference books or the Internet to find out about the **North** and **South Poles**.

   a How are they the same?   **b** How are they different?

Keeping in touch

Do you have friends or family in another country? How do you keep in touch with them?

All countries have a **postal service** to deliver letters. In many countries people have to collect their letters from the nearest post office. In other countries the letters are delivered to people's homes. What happens to the letters where you live?

When you send a letter, the price of the stamp you use depends on the weight of the letter and where it is going. How much would you have to pay to send a letter to your friend?

Letters travelling long distances go by air mail. Those going shorter distances travel by road, rail and, when necessary, by sea.

Rashid has written a letter to his cousin in Australia.

1 Rashid takes the letter to a post office. The man at the post office counter puts the correct stamp on it.

2 The letter goes to the main sorting office. Letters to places overseas are sorted by hand and put into bundles.

3 The bundle of letters for Australia is taken by van to the airport.

4 When the aircraft arrives in Australia the next day, the letters are unloaded into vans. They are taken to the local sorting office.

5 After the letters have been sorted into bundles for the local delivery areas, they are loaded onto small vans.

6 The local postman delivers the letter.

Rashid's letter had to travel a long way. Most letters have much shorter distances to travel, although the basic system is the same. As long as a letter is properly addressed and paid for, it can be sent anywhere in the world. It is only the postal service in the country from which the letter is sent that keeps the money you pay. Letters that arrive from that country are handled free of charge.

Did you know?

The first stamp was issued in the United Kingdom in 1840. It was called the Penny Black because of its cost and colour.

Activities

1 Work with a group of friends. Collect some old postcards. Read the dates and places on the postmarks. What kind of messages did people send? Can you find a way to display your results?

2 Collect four to six stamps from each of several countries. Stick the stamps from each country on a blank postcard or sheet of paper.

   a Find out where each country is on a world map.

   b Look carefully at each of the stamp collections. What do they tell you about each of the countries?

Electronic mail (e-mail)

Until about 40 years ago, the **postal service** was the only way for most people to send letters. Nowadays you can send a letter or short message much quicker using **e-mail** or a short message service (SMS).

Text messages (SMS)

You can now send a short message quickly and cheaply using a mobile phone. The **text message** is then received by the other person on his or her mobile phone.

Although text messages are usually short messages, you can add photographs, videos or sound content to them.

Text messaging is also commonly used for voting on television shows, for competitions, and by doctors, dentists and hairdressers to remind us that we have an appointment.

Dear Mr Baxter, You have an appointment on 17/08/14 at 08:25. Please contact the Dental Clinic if you are unable to attend.

E-mail messages

Have you sent a message by e-mail?

E-mail is an even faster way to send a letter or message. Information is sent from one computer to another using the telephone networks. It is also possible to send an e-mail message from mobile phones, electronic tablets and television sets.

As part of the e-mail network you can send and receive messages from all over the world almost instantly using a computer, a tablet or a smart phone.

Unlike text messaging, which is mainly used for short messages, e-mail can be used to send letters, or longer messages. Attachments can be added to the e-mail which may include scanned images, photos, videos, Word files and other larger documents.

Instant messaging

With instant messaging you can use the Internet to have a real-time text conversation with someone. This can be done using a smart phone, a tablet, an Internet TV or a computer. You can also chat with more than one person at the same time in an Internet chat room.

Linking the world

E-mail, text messaging and instant messaging technology is widely used by schools, colleges, businesses and people in their homes. This technology can link people not only in the same country but all over the world. In only a few minutes, you can find out what life is like in another country directly from people living there.

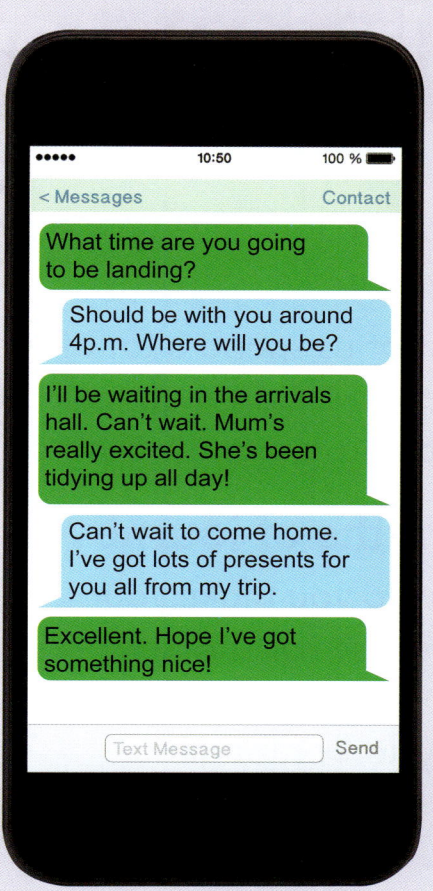

Did you know?

An American computer programmer called Raymond Tomlinson sent the world's first ever e-mail message (to himself) in 1971.

Activities

1 Work with a friend.

   a Talk about how you would send an urgent message to a friend.

   b Would you send a fax, a letter, a text message or an e-mail? Say why.

2 Write a simple description for a friend telling him or her how to send an e-mail message or a **text message**.

3 Write a message that you could send by e-mail to someone of your own age in one of the countries mentioned in this book.

Daily journeys

What is a journey? Why do we make journeys? What journeys do you make each day?

Most of us spend a lot of time moving from one place to another. Children go to school, adults go to work. People go shopping, go to the theatre, visit relatives or go on holiday.

Lengths of journey

How do you make your journeys? For shorter journeys we may walk or cycle. For longer journeys we may take a bus, a tram, the metro, a train or go by car. On very long journeys we may travel by plane. Nowadays, for example, it is quite easy to fly from Cairo to Cyprus, the United Arab Emirates or Jordan and back in a day.

People who travel to work in a large town or city are called **commuters**.

Commuters at a subway station in New York, USA, in the 'rush hour'.

Commuters and the 'rush hour'

Commuters make most of their daily journeys by road or rail. Many jobs start and end at about the same time each day, so people want to travel at the same time. Twice a day the buses, trams and trains are packed with people and the streets are crowded with **traffic**. We call these times 'rush hours'.

There may be smaller rush hours near your school when everyone arrives and leaves. It is not surprising that the smallest hold-up during the rush hour can cause chaos.

Rush-hour traffic in Dubai. Dubai has the busiest streets in the Middle East.

Millions of commuters travel by train each day in Japan.

Did you know?

Tokyo in Japan is probably the world's most crowded city. About 36 million people live in and around Tokyo, and many more commute to work there. Every 24 hours, 3 million people pass through Tokyo's Shinjuku Station, the world's busiest, where a train arrives every 3 seconds.

Activities

1 a Write a list of all the journeys you and your family make in a day, and in a week.

   b Draw a spider diagram to show how many different ways the people in your family make their journeys.

2 Work with a group of friends and talk about the jobs done by the adults you know. Find out where and how far the adults travel to work. Who has to make the longest journey?

5 What's in the news?

Weather in the news

What is the **weather** like where you are today? Is it hot or is it cold? Is it raining or is it dry? In some countries the weather is almost the same day after day. In many other countries, the weather changes from day to day and sometimes from hour to hour. It is not really surprising that the weather is often in the news.

Floods

What happens when it rains heavily for a long time, or when deep snow melts? Where does the water go?

Water from rain and melted snow runs into rivers. Sometimes there may be more water than a river can hold and the river **floods**. The water spills over the banks of the river and spreads across low-lying land nearby.

The sea and wind can also cause flooding. If a strong wind blows towards the land when there is a high tide, the land near the coast may be flooded.

In Bangladesh and some other countries, floods occur regularly. They destroy houses, **crops** and farmland, and wash away beaches. They can kill or injure people and animals.

Flooding caused when **Typhoon** Yolanda struck the Philippines in November 2013.

If there is an earthquake under the sea, or if a volcano erupts under the sea, the land along the coast can also be flooded.

Floods can cause great damage and loss of life. What damage has been caused by the floods in the pictures on page 36?

The weather and us

Why does the weather change? Weather changes are blown around the world by the wind. Winds are formed because the Sun warms some parts of the Earth more than others. The difference in **temperature** makes the air move.

We all need rain. It provides us with water for drinking, for washing ourselves and our clothes, for growing crops and for keeping animals alive. Without rain we would die.

River floods are also sometimes useful. When a river floods, it spreads mud and sand over the surrounding land. This mud and sand makes very **fertile** soil for growing crops.

Did you know?

In 1887, one million people were killed when the Hwang Ho River in China flooded.

Activities

1 Some newspapers and websites show how much rain falls each day in the larger towns and cities of the world.

   a Choose a country, collect these figures for a week and record them.

   b Write a list of the places from your country with the wettest town or city at the top and the driest at the bottom.

   c Use an **atlas** to find out where these towns and cities are.

2 Find a picture of a weather event such as flooding or heavy snow somewhere in the world. Pretend you are a newspaper reporter. Write a news report about what happened before, during and after the flooding or heavy snow.

How can we find out what the **weather** will be like tomorrow? Why do we need to know?

Scientists try to predict, or forecast, what the weather will be like over the next few days. This helps us to plan our lives.

Weather forecasts

The **weather forecast** is important to many people in their work, particularly those who work outdoors. Farmers need help to decide when to plough their land and sow and harvest their **crops**. The captain of a ship might not want to set sail if there was going to be a bad storm.

A weather forecast includes information about **temperature**, rainfall, wind, cloud and sunshine. **Symbols** are used on the map to show the weather in different places. What do the symbols on this map mean?

Every day, all over the world, hundreds of weather forecasts are given out on television and radio. They also appear in newspapers and on websites. Some telephone companies give out forecasts. Weather forecasts are given for continents, countries or local areas.

A weather satellite.

A weather balloon contains instruments that measure the weather conditions high in the sky.

Weather information

The information to produce the weather forecasts comes from ships, aircraft, weather balloons and weather stations all over the world, as well as weather satellites in space. Nowadays, satellites continuously photograph the Earth's surface from space. Every patch of cloud and every storm can be seen as it forms and changes.

Small weather stations, like this, record lots of details about the local weather.

Weather scientists, called **meteorologists**, analyse all the information about the weather using computers. By looking at all this information, they can make predictions, or forecasts, of what the weather is likely to do next.

Weather stations

Altogether there are about 11 000 weather stations all over the world. The information from each weather station is passed to regional centres, such as the local airport, and then on to one of the main meteorological offices in the country.

Activities

1. Record the weather every morning and afternoon for a week. Use symbols to show what the weather is like. Does the weather change much where you live?

2. Look at newspapers or use the Internet to find out what the temperatures are in Abu Dhabi, Berlin, Cairo, London, Singapore and New York. Do this each day for a month and record your results.

   a Which city has the warmest temperatures?

   b Which city has the coolest temperatures?

   c Repeat this activity at a different time of the year. How do the results compare?

Local changes

Two hundred years ago most people lived in villages or on **farms** in the countryside. They grew their own food in their gardens and fields. Most of their journeys were made on foot, or by using animals such as horses or camels. How is life different today?

Changing settlements

Today there are millions more people in the world. More and more of them live in towns and cities. They travel by car, bus, train or airplane. They buy their food in supermarkets. Villages are changing as new houses are built and the growing number of motor vehicles is causing problems.

Can you tell which buildings are the newest? How do you know?

Planning for change

In some countries, new buildings are simply added to a village, town or city but in many countries people must get permission from a government department to put up a new building.

Planning officers make sure that historical buildings are not damaged.

Planning officers control changes in the local villages, towns or cities. They check that buildings blend in with the local scene and do not badly affect the neighbours or the **environment**. They make sure that new buildings are put in the correct area for that type of development. Certain planning officers also control the building of new roads or the digging of mines, quarries and pits.

Finding out about local changes

What changes are taking place in your local area? In some places, details of planning applications for new buildings are published in the local newspaper. In other places, lists are put up in the local public library or some other official building. The local newspaper, radio or television station will probably describe any big changes that are planned.

Sometimes there is an exhibition showing models of what new buildings or roads will look like.

Did you know?

Today more than half the people in the world live in cities. People move to cities from the countryside hoping to find work, earn money and make a better life for themselves and their families.

Activities

1 Find out about a change that is taking place in your area, such as the building of a new road or apartment block, or the closing down of a factory. Imagine you are a newspaper reporter. Write a news story for your newspaper about this change. Draw a map with pictures to show where the change is taking place.

2 Collect pictures of your village, town or city. What are the oldest pictures you can find? How has your village, town or city changed? Can you find any early plans or maps? As a class, display your pictures, plans and maps.

3 Imagine you have to plan a new village, town or city centre. What sort of buildings will it need? Draw a picture to show what your new village, town or city will look like.

Traffic news

Is **traffic** a problem in your area?

The number of motor vehicles is growing, and in 50 years' time there may be twice as many as there are now. Already in many places the roads and streets are clogged with traffic.

Motor vehicles and people

If there were no cars, we would have much less choice in where to live, work or spend our leisure time. If there were no trucks, we would have much less food and fewer other goods in our shops, and these items would be much more expensive.

The motor vehicles on our roads give out exhaust gases that are harmful to people and damage buildings. They also affect the world's **climate** by making it warmer.

Traffic jams are annoying to drivers and damaging to people's health and the **environment**.

Traffic jams

A heavy shower of rain or snow can make the roads slippery and dangerous. Fog or a dust storm can make it difficult for drivers to see where they are going.

Air **pollution** caused by motor vehicles in Mexico City. Air pollution causes at least 150 000 deaths worldwide each year.

Many different **weather** conditions can cause traffic jams. Repairs to roads and road accidents can also cause traffic jams. Even a small accident can block a road for a time, causing the traffic to slow down and stop.

A serious road accident can block the road for several hours.

Planning a road journey

Imagine you are planning a journey by road. Where can you find out about traffic delays?

It is a good idea to listen to the traffic reports on the radio or look at local travel reports on the Internet. These give up-to-date information on accidents, road repairs, bad weather conditions and other things that make road travel difficult.

Did you know?

Moscow, the **capital** city of Russia, has the worst traffic jams in the world. In December 2012, a snowstorm caused a traffic jam 190 kilometres long. Some motorists were trapped in their cars for up to three days. Moscow also has one of the world's highest numbers of road accidents.

Activities

1   Write a list of the kinds of things that might delay road traffic and railway trains. Which kind of traffic is most easily delayed?

2   Some people say that if we had bigger and better roads, then road traffic delays would not occur so often. Talk about this with a friend. Do you agree or disagree? Write down the arguments for and against bigger and better roads.

Map of the World

NORWAY

UNITED
KINGDOM

AUSTRIA
SWITZERLAND

ROMANIA
BULGARIA

FRANCE

PORTUGAL
SPAIN

ITALY

TURKEY

CYPRUS

IRAQ
JORDAN

IRAN

RUSSIA

CHINA

JAPAN

LIBYA

EGYPT

SAUDI
ARABIA

UNITED ARAB
EMIRATES

INDIA

BHUTAN

BANGLADESH

BURMA

THAILAND

PHILIPPINES

VIETNAM

MALI

CHAD

SUDAN

SRI LANKA

DEMOCRATIC
REPUBLIC
OF CONGO

TANZANIA

ZAMBIA

AUSTRALIA

BOTSWANA

NEW
ZEALAND

ANTARCTICA

Glossary

Aerial photograph A photograph taken from above, usually from an airplane, balloon or a drone.

Atlas A book of maps.

Capital The most important city in a country.

Climate The typical weather of a place over the whole year.

Commuter Someone who regularly travels to work, especially by bus or train.

Compass A device for showing which direction is which; the needle always points north.

Crop A plant grown by farmers.

Electronic mail (e-mail) The use of a computer system to send or receive messages over long distances.

Endangered An animal or plant that is in danger of dying out.

Environment Your surroundings.

Equator An imaginary line round the Earth at an equal distance from the North and South Poles.

Extinct Not existing anymore.

Farm The land and buildings used for growing crops and looking after animals.

Fertile Able to produce good crops.

Fertiliser A chemical put into the soil to make crops grow better.

Fuel Anything that can be burned to produce heat or light.

Globe A ball with a map of the whole world on it.

Gorge A narrow, deep valley with steep sides.

Hay Dried grass used for feeding farm animals.

Irrigation To put water onto the land so that crops grow well.

Key The key to a map tells you what the various symbols on it mean.

Meadow An area of natural grassland, often used for grazing animals.

Meteorologist A scientist who studies the weather.

Minerals Useful chemical substances in the ground obtained by mining or quarrying.

Monsoon A strong wind in and around the Indian Ocean that brings heavy rain in summer.

Mountain A very high part of the Earth's surface.

North Pole The most northerly part of the Earth.

Oasis A place where water can be found in the desert.

Pesticide A chemical sprayed onto crops to stop insects and other pests from destroying them.

Pollute/Pollution When substances such as air, water or the soil are spoiled or made dirty by people.

Postal service The service that carries letters and parcels.

Scale The scale on a map tells us how much the distance on paper (e.g. I centimetre) is actually worth on the ground (e.g. 50 kilometres).

Season A time of the year when you can expect a special type of weather.

Settlement A place where people live. Villages, towns and cities are settlements.

Source The place where a river starts.

South Pole The most southerly part of the Earth.

Spring A place where water naturally flows out of the ground.

Symbol A small, simple drawing used on a map to show where things such as car parks and castles are.

Temperature How hot or cold someone or something is.

Text message A short message sent using a mobile phone.

Tourist Someone visiting a place for pleasure rather than for work.

Traffic Cars, buses, trucks, bicycles, and so on, travelling along a road.

Typhoon A huge storm that sweeps into eastern Asia from the Pacific Ocean.

Valley A line of low land between hills or mountains.

Weather The rain, wind, snow, sunshine and temperature at a particular time or place.

Weather forecast A prediction of what the weather is going to be like.